FLOWERS WE EAT

Katherine Rawson

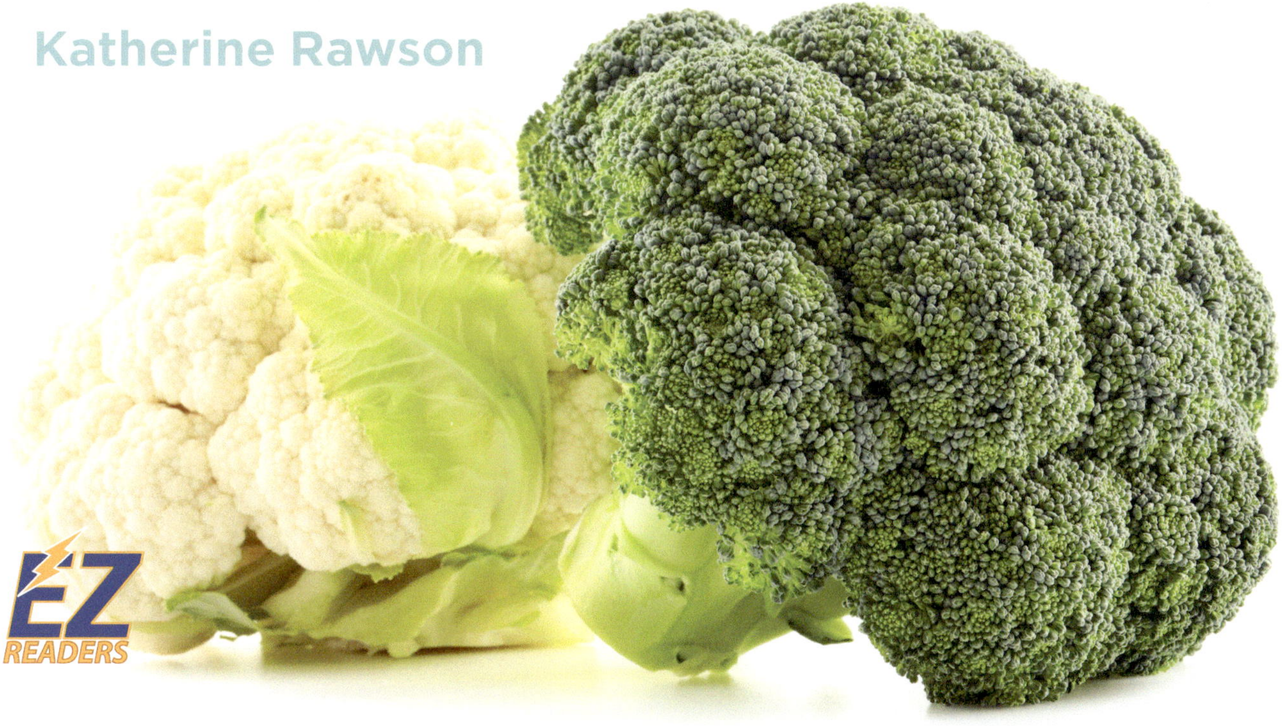

EZ READERS

Creating Young Nonfiction Readers

EZ Readers lets children delve into nonfiction at beginning reading levels. Young readers are introduced to new concepts, facts, ideas, and vocabulary.

Tips for Reading Nonfiction with Beginning Readers

Talk about Nonfiction
Begin by explaining that nonfiction books give us information that is true. The book will be organized around a specific topic or idea, and we may learn new facts through reading.

Look at the Parts
Most nonfiction books have helpful features. Our *EZ Readers* include a Contents page, an index, and color photographs. Share the purpose of these features with your reader.

Contents
Located at the front of a book, the Contents displays a list of the big ideas within the book and where to find them.

Index
An index is an alphabetical list of topics and the page numbers where they are found.

Photos/Charts
A lot of information can be found by "reading" the charts and photos found within nonfiction text. Help your reader learn more about the different ways information can be displayed.

With a little help and guidance about reading nonfiction, you can feel good about introducing a young reader to the world of *EZ Readers* nonfiction books.

Mitchell Lane
PUBLISHERS
2001 SW 31st Avenue
Hallandale, FL 33009
www.mitchelllane.com

First Edition, 2021.

Author: Katherine Rawson
Designer: Ed Morgan
Editor: Morgan Brody

Names/credits:
Title: Flowers We Eat / by Katherine Rawson
Description: Hallandale, FL :
Mitchell Lane Publishers, [2021]

Series: Plant Parts We Eat
Library bound ISBN: 978-1-58415-043-5
eBook ISBN: 978-1-58415-047-3

EZ Readers is an imprint of Mitchell Lane Publishers.

Photo credits: Freepik.com, Shutterstock

Contents

Plants have flowers. Flowers begin as tiny **buds**. When the buds open, they become flowers.

Flowers make **pollen**.
They use the pollen to make seeds.
The seeds will grow into new plants.

Some flower buds are good to eat.
Broccoli buds are good to eat.

Did You Know?

The top part of the broccoli plant is called the crown.

Did You Know?

Purple broccoli and cauliflower turn green when cooked.

Broccoli is usually green.
Sometimes it is purple.

Broccoli tastes good **raw**.
It tastes good with **dips** and in salads.

Broccoli tastes good cooked.
It tastes good in soups.
It tastes good on pizza.

Did You Know?

If you don't pick broccoli on time, the buds will open and become yellow flowers.

Did You Know?

Sunlight can turn cauliflower light green or yellow. Cauliflower stays white because the leaves cover the flower and keep the sunlight away.

Cauliflower is good to eat.
Cauliflower is usually white.
Sometimes it is orange or purple.
Sometimes it is green.

Cauliflower tastes good with dips and in salads.
It tastes good in soups.
It tastes good **roasted** in the oven.

Broccoli and cauliflower are delicious flowers we can eat.

Glossary

bud
The part of the plant that becomes a flower

dips
A sauce that food is dipped into

pollen
Yellow powder used by a flower to make seeds

raw
Not cooked

roasted
Cooked in the oven

Sources

https://www.britannica.com/plant/broccoli https://www.britannica.com/plant/cauliflower

https://cals.arizona.edu/fps/sites/cals.arizona.edu.fps/files/cotw/Cauliflower.pdf

https://homeguides.sfgate.com/keep-cauliflower-white-during-growing- https://hortnews.extension.iastate.edu/2008/4-9/cauliflower.html

https://www.csmonitor.com/The-Culture/Gardening/2009/0909/the-science-behind-purple-beans

https://www.sciencedirect.com/science/article/pii/S0960982295000728

http://www.mbgnet.net/bioplants/pollination.html

Further Reading

Web Pages:
Read more about flowers:
https://extension.illinois.edu/gpe/case1/c1facts2d.html
https://www.dkfindout.com/us/animals-and-nature/plants/parts-flower/

Watch a video about how a plant grows:
https://vermont.pbslearningmedia.org/resource/evscps.sci.life.seed/from-seed-to-fruit/

Read and watch a video about pollination:
https://kidsgrowingstrong.org/pollination/

Books:
Pollination
By Dona Herwick Rice
(Teacher Created Materials, 2014)

Seed to Plant
by Kristin Baird Rattini
(National Geographic Children's Books, 2014)

Index

About the Author

Katherine Rawson loves growing, cooking, and eating vegetables. She also loves writing. So, she thought it would be a great Idea to write books about plants we eat. She loves eating both broccoli and cauliflower and is always looking for new ways to cook these flowers.